International Environmental Labelling

Vol.10 of 11
For All People who wish to take care of Climate Change,
Financial Products & Services:
(Banking, Professional Advisory, Wealth Management, Mutual
Funds, Insurance, Stock Market, Treasury/Debt Instruments,
Tax/Audit Consulting, Capital Restructuring,
Portfolio Management)

Jahangir Asadi

Vancouver, BC CANADA

Suggest an ecolabel

If you think that we missed a label and/or you are an ecolabelling body, please consider to submit for the next editions of our 11 Volumes International Eco-labelling Book series. Please send your details, and we'll review your suggestions. Our goal is to be as comprehensive as possible, so thank you for your help!
info@TopTenAward.Net

Published by: Top Ten Award International Network
Vancouver, BC **CANADA**
Email: Info@TopTenAward.net
www.TopTenAward.net

Ordering Information:
Quantity sales. Special discounts are available on quantity purchases by universities, schools, corporations, associations, and others. For details, contact the "Sales Department" at the above mentioned email address.

International Environmental Labelling Vol.10/J.Asadi—2nd ed.
ISBN 978-1-7775268-1-8

Contents

About TTAIN 10

Introduction 13

General principles of environmental labelling 20

Types of environmental labelling 24

Types I environmental labelling 28

Types II environmental labelling.................................... 46

Types III environmental labelling 52

All about 'Eco-friendly' Financial Products and/or Service........... 54

Green Retail Banking... 57

Green Corporate... 58

Eco Friendly Insurance 60

Asset Management... 61

Are Crypto Currencies Inherently Bad............................... 61

TTAIN Pioneers.. 72

Bibliography 79

Search by logos 86

Federal Investment in Algae 90

Environmental friendly photos 92

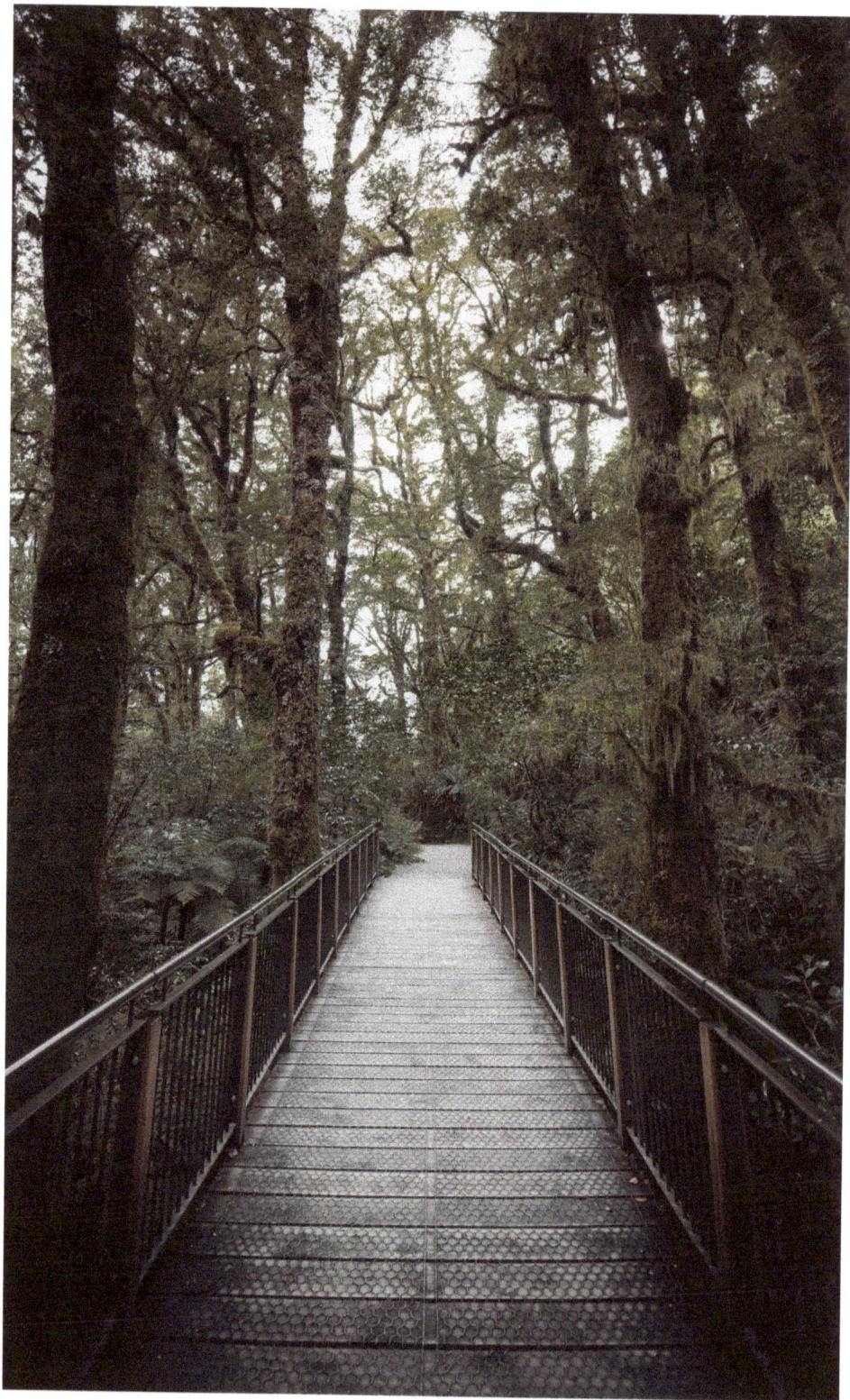

I dedicate this book to my son, Taha

We hope that, 10,000 years from now, future generations will be able to see flowers that provide bees with nectar and pollen and... BEES provide flowers with the means to reproduce by spreading pollen from flower to flower...

Jahangir Asadi

Acknowledgements:

I wish to thank my committee members, who were more than generous with their expertise and precious time. I would like to acknowledge and thank the Top Ten Award International Network for allowing me to conduct my research and providing any assistance requested.

It should be noted that all the required permissions for using the logos and trade marks has been obtained to be published in this volume.

Why do we need to use environment friendly?
Since the world is corrupted with pollution and toxic amount of materials, making it sustainable can be a good call.
Going eco-friendly also improves your quality of life in terms of mortality, age, diseases etc.
You might have a better shot at living a quality life with health if you chose to go eco-friendly.

ABOUT TTAIN

Top Ten Award International Network

Top Ten Award international Network (TTAIN) was established in 2012 to recognize outstanding individuals, groups, companies, organizations representing the best in the public works profession.

TTAIN publishing books related to international Eco-labeling plans to increase public knowledge in purchasing based on the environmental impacts of products.

Top Ten Award International Network provides A to Z book publishing services and distribution to over 39,000 booksellers worldwide, including Apple, Amazon, Barnes & Noble, Indigo, Google Play Books, and many more.

Our services including: editing, design, distribution, marketing TTAIN Book publishing are in the following categories:

Student
Standard
Business
Professional
Honorary

We focus on quality, environmental & food safety management systems , as well as environmnetal sustain for future kids. TTAIN also provide complete consulting services for QMS, EMS, FSMS, HACCP and Ecolabeling based on international standards.

ISO 14024 establishes the principles and procedures for developing Type I environmental labelling programmes, including the selection of product categories, product environmental criteria and product function characteristics, and for assessing and demonstrating compliance. ISO 14024 also establishes the certification procedures for awarding the label.

TTAIN has enough experiences to help create new ecolabeling programmes in different countries all over the world.
For more detail visit our website : http://toptenaward.net
and/or send your enquiery to the following email:
info@toptenaward.net

CHAPTER 1

Introduction

This book is dedicated to the subject of environmental labels. The basis for the classification of its parts goes back to the types of environmental labelling according to the classifications provided by the International Organization for Standardization. In each section, while presenting the relevant definitions, I mention the existing international standards and present examples related to each type of labelling. Environmental labelling is an important and significant topic, and its richness is added to every day, which has attracted the attention of many experts and researchers around the world. The idea of compiling this book, came to my mind when I observed that national environmental labelling models have been developed in most countries of the world, but in many other countries, the initial steps have not been taken yet. Therefore, I decided to create the first spark for the development of environmental labelling patterns in other countries by collecting appropriate materials and inserting samples of labelling patterns of different countries of the world. It should be noted that the description of each environmental label in this book does not indicate their approval or denial; they are included only to increase the awareness of all enthusiasts and consumers of the meanings and concepts derived from such labels. We hereby ask all interested parties around the world who wish to start an environmental labelling program in their country to benefit from our intellectual assistance and support in the form of consulting contracts. Increasing human awareness of the urgent need to protect the environment has led to changes in all levels of activities, including the production of marketing products, consumption, use, and sale of goods and services at the national and international levels. Stakeholders involved in environmental protection include consumers, producers, traders, scientific and technological institutes, national authorities, local and international organizations, environmental gatherings, and human society in general. Decisions by consumers and sellers of products are made not only on the basis of key points such as quality, price, and availability of

products but also on the environmental consequences of products, including the consequences that a product can have before, after and during production. The most important environmental consequences include water, soil, and air pollution along with waste generation, especially hazardous waste. Further consequences include noise, odor, dust, vibration, and heat dissipation as well as energy consumption using water, land, fuel, wood, and other natural resources. There are further effects on certain parts of the ecosystem and the environment. In addition, the environmental consequences not only include the natural use of the products but also abnormal and even emergency or accidental uses. The basis of studies and studies in this field is done through product life cycle evaluation, which generally involves the study and evaluation of environmental aspects and consequences of a category (product, service, etc.) because of the preparation of raw materials for production until they are used or discarded. Sometimes the phrase "review from cradle to grave" is used for such an evaluation. In addition to the above, the environmental consequences that may occur at any stage of the product life cycle, including the preliminary stages and its preparation, production, distribution, operation, and sale, should also be considered when evaluating it. This type of evaluation refers to product life cycle analysis from an environmental point of view," which is a useful tool for measuring the degree of environmental health

of a product, comparing different products, improving product quality, and confirming the environmental health claims of the product. The environmental health analysis tool for products and services facilitates their placement in domestic or foreign markets, considering that the awareness of consumers and retailers about the environmental consequences of the product has increased, as has the accurate and explicit measurement by the people in charge at all levels. Local, national, and international in the field of environmental protection. Products that can claim to be environmentally complete in all stages of their life cycle and meet the mandatory and optional environmental needs are considered successful products.

Environmental messages refer to the policies, goals, and skills of product manufacturing companies as part of the environmental management systems in which they are applied, and consumers and retailers are increasingly paying attention to this issue when making purchasing decisions. In addition, companies have been encouraged and even forced to adapt their environmental management systems to agencies and retailers and to local, national, international, and other environmental issues.

The environmental health message of a product can be conveyed to the consumer in various ways, including implicitly or explicitly. For example, the implicit or implicit message conveyed directly by the product to the customer is that the product is suitable for the intended use and purpose, and, without material waste in size, weight, and dimensions, is perfectly proportioned and without additional packaging. Sometimes it is necessary to convey these messages and claims about the correctness of the product quite clearly through magazines or other media as well as through certificates that are accurate, simple, and convincing to the consumer in the form of a label. These messages must be accurate and fact-based; otherwise they will nullify the product and create contradictory effects. Confirmation of these claims by a third-party organization will increase its credibility. It should also be noted that the multiplicity of these messages, depending on the type of products or companies producing them, confuses consumers in the market and also creates artificial boundaries or causes a differentiated distinction against certain products or companies. Various models, principles, and methods have been provided by local, regional, national, and international organizations to demonstrate product life cycle analysis and other guidelines on environmental management systems and their labels. At the national level, significant advances have been made in the design of environmental labels in various countries, including developing countries and the Scandinavian countries. For example, the first project was designated in Germany as a Blue Angel in 1977, later on Canada in 1988, the Scandinavian countries and Japan in 1989, the United States and New Zealand in 1990, India, Austria, and Australia in 1991, And in 1992, Singapore, the Republic of Korea, and the Netherlands developed their national environmental labelling. Environmental labels are an environmental management tool that is the subject of a series of ISO 14000 standards. These environmental labels provide information about a product or commodity in terms of its broad environmental characteristics, whether it is about a specific environmental issue or about other characteristics and topics.Interested and pro-environmental buyers can use this information when choosing products or goods. Product makers with these environmental labels hope to influence people's purchasing decisions. If these environmental labels have this effect, the share of the product in question can increase, and other suppliers may create healthy environmental competition by improving the environmental aspects of their products and commodities. The overall goal of environmental labels is to convey acceptable and accurate information that is in no way misleading regarding the environmental aspects of products and commodities, and they encourage the consumer to buy and produce products that reduce stress on the environment. Environmental labelling must follow the general principles that the International Organization for Standardization has published in a collection entitled the ISO 14020 standard, which refers to these general principles here. It should be noted that other documents and laws in this field are considered if they are in accordance with the principles set out in ISO 14020.

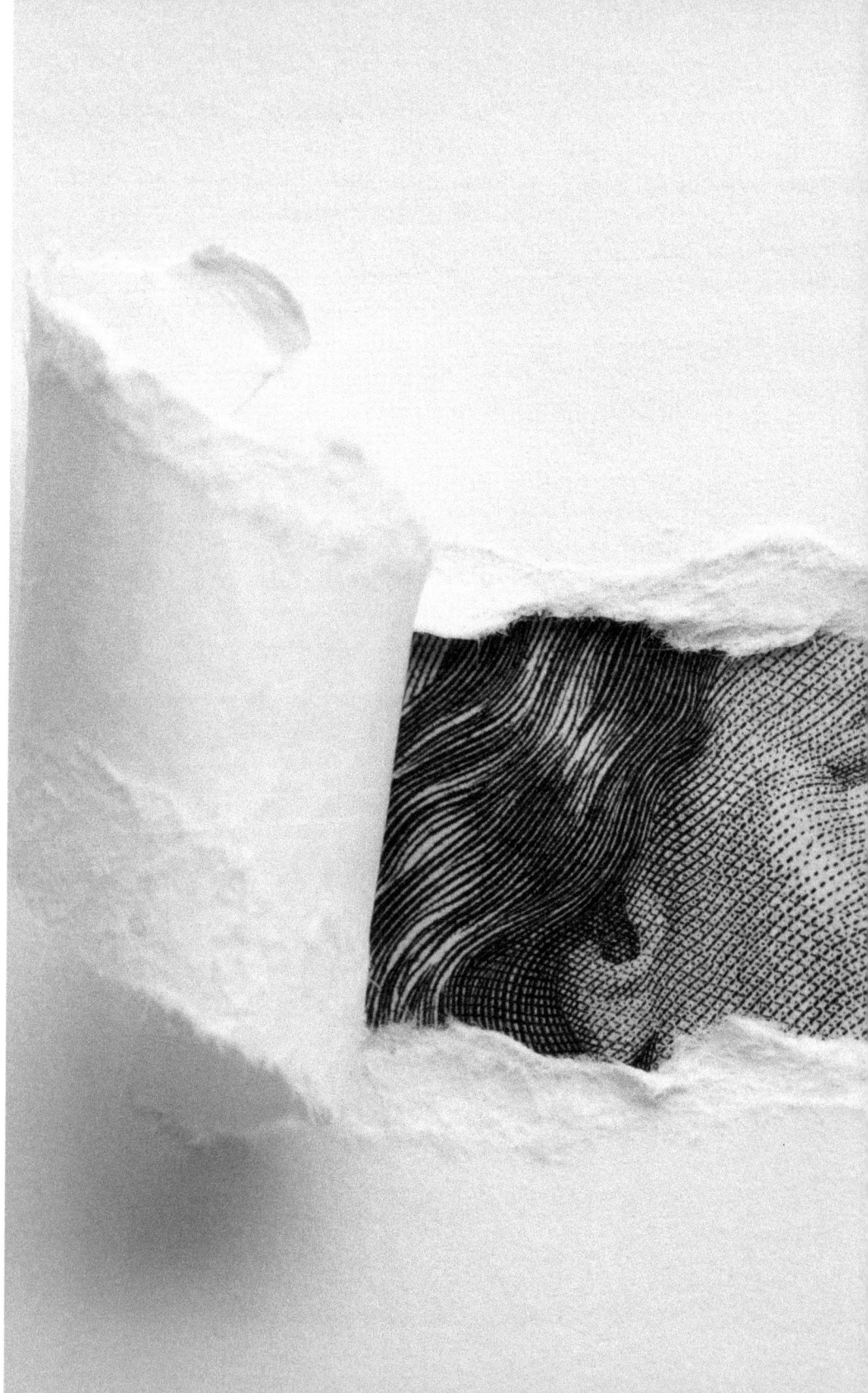

CHAPTER 2

General Principles on Environmental Labelling

1 The First Principle: Evironmental notices and labels must be accurate, verifiable, relevant, and in no way misleading and/or deceptive.

2 The Second Principle: Procedures and requirements for environmental labels will not be ready for selection unless they are implemented by affecting or eliminating unnecessary barriers to international trade.

3 The Third Principle: Environmental notices and labels will be based on scientific analysis that is sufficiently broad and comprehensive, and to support this claim, the product must be reliable and reproducible.

4 The Fourth Principle: The process, methodology, and any criteria required to support the announcements on environmental labels will be available upon request all interested groups.

5 The Fifth Principle: Development and improvement of environmental notices and labels should be considered in all aspects related to the service life of the product.

6 The Sixth Principle: Announcements on environmental labels will not prevent initiative and innovation but will be important in maintaining environmental implementation.

7 The Seventh Principle: Any enforcement request or information requirement related to environmental notices and labels should be limited to the necessary information to establish compliance with an acceptable standard and based on the notification standards and environmental labels.

8 The Eighth Principle: The process of improving the announcement and environmental labels should be done by an open solution with interested groups. Reasonable impressions must be made to reach a consensus through this process.

9 The Ninth Principle: Information on the environmental aspects of the product and goods related to an advertisement and environmental label will be prepared for buyers and interested buyers from a group consisting of an advertisement and an environmental label.

Cryptocurrencies?

However, co-locating bitcoin mining operations with zero-carbon resources, such as nuclear, hydro, wind and solar, could help reduce the carbon emissions associated with the mining itself. Co-location could also give a financial boost to power plants that might be able to sell their electricity at a higher price to miners instead of to the grid when demand and prices are low. This type of hybrid power plant/mine might even make uneconomical projects economical.

CHAPTER 3

Types of Environmental Labelling

At present, according to the classification provided by the International Organization for Standardization, there are three types of environmental labelling patterns:

1 Type I labelling: This labelling is known as eco-labelling, and because it is difficult to translate this word into many languages, it presents another reason to adhere to a numerical classification system. In the content of Type I labelling, a set of social commitments that creates criteria according to the scientific principles on the basis of which a product is environmentally preferable is discussed. Consumers are then instructed in assessing environmental claims and must decide which packaging is more important.

2 Type II labelling: refers to the claims made on product labels in connection with business centers. This includes familiar claims such as recyclable, ozone-friendly, 60% phosphate-free, and the like. This type of labelling can be in the form of a mark or sentence on the product packaging. Some of them are valid environmental claims—and some can be completely misleading. Usually, all countries have laws against deceptive advertisements, so why has the International Organization for Standardization discussed this issue? The answer is that it is not clear whether the environmental claims have a technical basis or whether the ad is meaningless.

3 Type III labelling: is a distinct form of third-party environmental labelling pattern designed to avoid the difficulties that can result from type-one labelling. Technical committee for Environment of International organization for Standardization has undertaken a new project to standardize guidelines and Type III labelling methods. One of the main objections raised by industries to Type I labelling is the basis for its management.

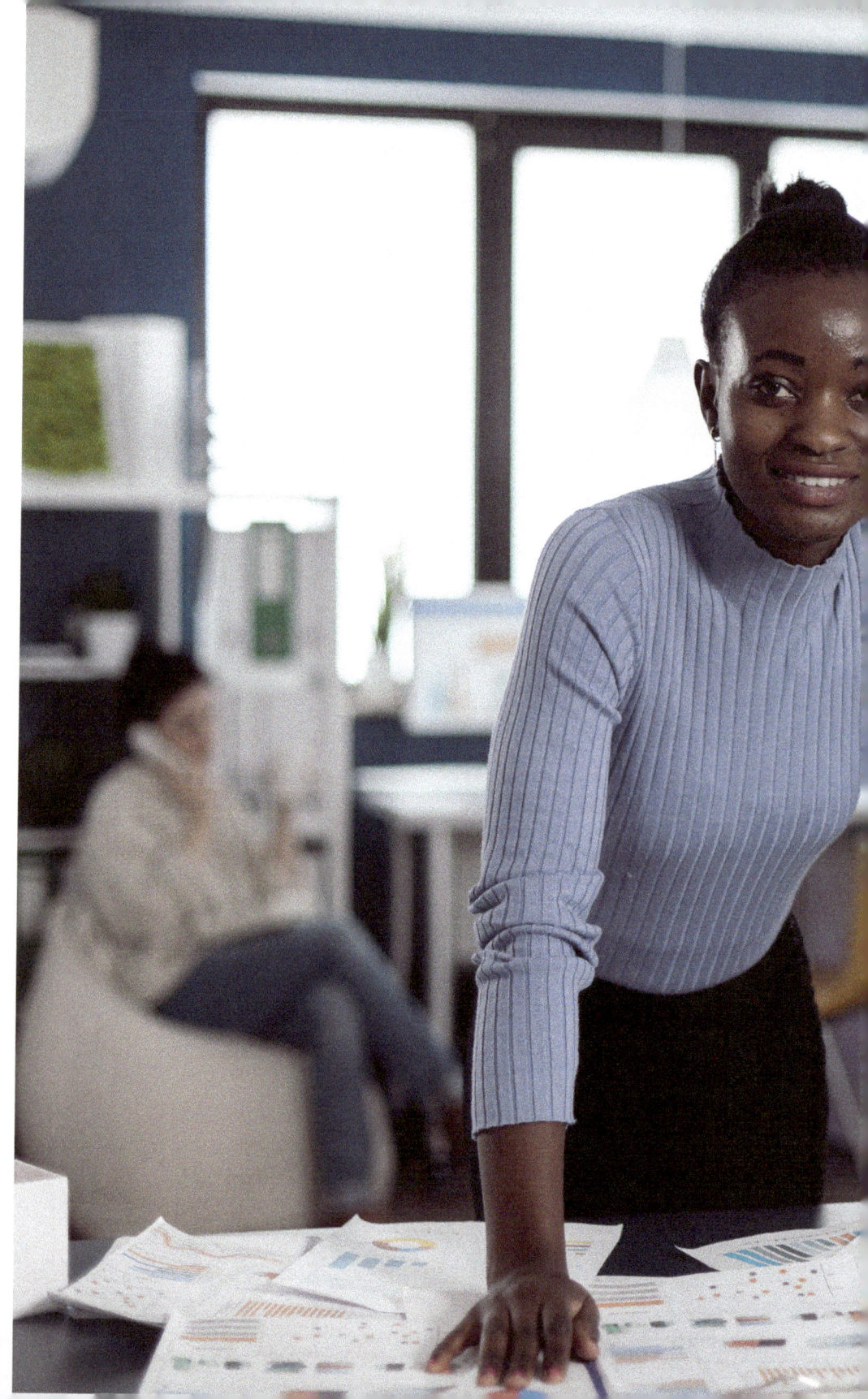

Which is the greenest cryptocurrency?
TRG Datacenters report suggests that Dogecoin is one of the most environmentally friendly cryptocurrencies out there. Its research suggests that the cryptocurrency consumes only 0.12 kWh of energy per transaction, compared to 707 for Bitcoin.

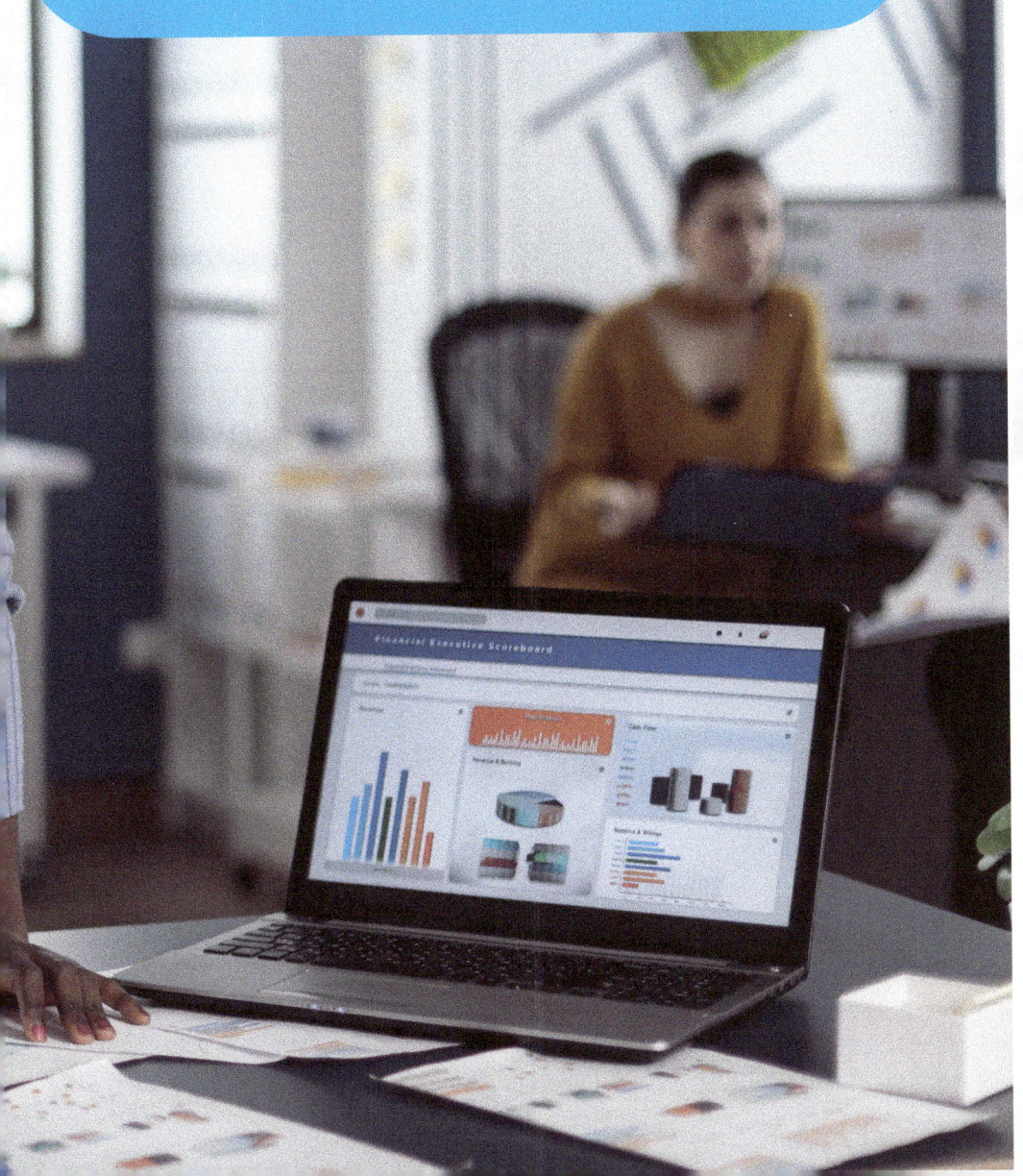

CHAPTER 4

Type I Environmental Labelling

Type I labelling: This labelling is known as eco-labelling, and because it is difficult to translate this word into many languages, it presents another reason to adhere to a numerical classification system. In the content of Type I labelling, a set of social commitments that creates criteria according to the scientific principles on the basis of which a product is environmentally preferable is discussed. Consumers are then instructed in assessing environmental claims and must decide which packaging is more important.

Type I adhesive has the following specifications:
A. Has an optional third-party template.
B. When the product meets a certain standard, the labelling of this product is included.
C. The purpose of this program is to identify and promote products that play a pioneering role in terms of environment, which means its criteria are at a higher level than the average environmental performance.
D. Acceptance/rejection criteria are determined for each group of products and are publicly available.
E. The criteria are adjusted after considering the environmental consequences of the product life cycle.

Examples of Type I Labelling:
In this section, and considering the importance of this type of labelling, I provide a description of some examples of Type I labelling related to some countries along with a   list of products on which this mark is placed.

Germany

FSC® is a global not-for-profit organization that sets the standards for responsibly managed forests, both environmentally and socially. When timber leaves an FSC certified forest they ensure companies along the supply chain meet our best practice standards also, so that when a product bears the FSC logo, you can be sure it's been made from responsible sources. In this way, FSC certification helps forests remain thriving environments for generations to come, by helping you make ethical and responsible choices at your local supermarket, bookstore, furniture retailer, and beyond. www.fsc.org

FSC® International
Adenauerallee 134
53113 Bonn
E-mail: info@fsc.org
Phone: +49 (0) 228 367 66

FSC Canada
50 rue Sainte-Catherine Ouest,
bureau 380B, Montreal, QC H2X 3V4
Email: info@ca.fsc.org
Telephone: 514-394-1137

EKOagros

ORGANIC CERTIFICATION

Lithuania

EKOAGROS is the only institution in Lithuania for more than 20 years carrying out certification and control activities of organic production and products of national quality, also providing services of certification activities in accordance with the foreign national and private standards in foreign countries. From year 2017 EKOAGROS is accredited as certifying agent to conduct certification activities on crops, wild crops, livestock and handling operations in accordance with USDA NOP.

Contact information:
EKOAGROS
Address K. Donelaicio str. 33, LT-44240 Kaunas, Lithuania
Tel. No. +370 37 20 31 81
Website: www.ekoagros.lt

Kazakhistan

The ecological brand «ECO» is promoted in the country by the International Academy of Ecology which is the only leading non-profit organization in Kazakhstan on the market for the development of the industry of environmentally friendly products, services and processes.

Program for sustainable "ECO" development with the aim of providing comprehensive information on hazardous substances, environmental impacts and end-of-life guidance has been developed by the Academy. The Program demonstrates expression of our inner conviction that ambitious environmental principles should be included in all our proposals. It also coincides with the most frequent request from customers and business partners in many regions and different market segments.

The International Academy of Ecology cooperates with a number of reputable organizations in Kazakhstan, implements joint global projects, participates in environmental programs and events, engages in environmental education and promotes sustainable production and consumption. Trainings on the development of environmental labeling in different countries of the world are held.

Website: www.eko-kaz.kz
Email: mae_astana@mai.ru
Phone: +7 7172 57 43 68.

China

China Environmental United Certification Center (CEC), approved by the Ministry of Ecology and Environment of the People's Republic of China (MEE) and accredited by Certification and Accreditation Administration Committee of PRC, is a comprehensive certification and service institution leading in environmental protection, energy saving and low carbon areas. . CEC is committed to serve building national ecological civilization; and has carried out research on environmental protection, energy saving, low carbon development strategies and solutions; has been continuously improving and innovating green industry evaluation system on industrial green development and transition CEC is building a bridge between green production and green consumption by offering independent, impartial and high-quality evaluation and certification service for government, enterprises and the public. CEC is a state-owned, non-profit, legal entity of independent third-party certification. It integrates the certification resource from the former National Accreditation Center for Environmental Conformity Assessment, the Secretariat of China Environmental Labelling Products Certification Committee, Environmental Development Center of MEE, the Chinese Research Academy of Environmental Sciences and other institutions. Business areas includes: products certification, management systems certification, services certification, addressing climate change, energy-saving and energy efficiency certification, green supply chain assessment, environmental stewardship, green credit assessment and green manufacturing system evaluation. CEC also carries out standard establishment and research project and international cooperation and exchanges, etc.

Contact:
Website: http://en.mepcec.com/
E-mail: zhangxiaoh@mepcec.com , zhangxiaoh@mepcec.com

Sri Lanka

Sri Lanka

National Cleaner Production Centre (NCPC), Sri Lanka was set up by UNIDO in 2002, as a project under the Ministry of Industry to provide the technical expertise and support to the industry and business enterprises in order to prevent pollution and conserve resources by the application of Cleaner Production (CP) and other proactive environmental management tools. NCPC Sri Lanka is registered as a Company by Guarantee not for profit organization under the Act No. 7 of 2007. Over the past two decades, it has evolved as the foremost sustainability solution provider in the country.

The ISO 9001:2015 certified Centre is a registered Energy Service Company (ESCO) under Sustainable Energy Authority (SEA) and a registered consultant under Central Environmental Authority (CEA). It is a founding member of UNIDO/UNEP Resource Efficient and Cleaner Production Network (RECP Net), a global family of 52 NCPCs. NCPC Sri Lanka is a member of Climate Technology Centre & Network (CTCN) and associate member of Global Eco-labelling Network (GEN). Accordingly, we at National Cleaner Production Centre (NCPC), Sri Lanka has developed Eco Labelling scheme under the ISO 14024:2018 - Environmental labels and declarations. NCPC Eco labelling scheme developed, with the Support of United Nations Environment Programme, Under One Planet Network Consumer Information Programme for Sustainable Consumption and Production (CI-SCP).

Contact:
Tel: +94 11 2822272/3,
Fax: +94 11 2822274
E mail: info@ncpcsrilanka.org
Web: www.ncpcsrilanka.org

Hong Kong

The Green Council is a non-profit, tax-exempt charitable environmental stewardship organisation and certification body (Reg. No.: HKCAS-027) of Hong Kong established in 2000. A group of individuals from different sectors of industry and academics shared the vision to help build Hong Kong into a world-class green city for the future. They formed the Green Council with the aim of encouraging the commercial and industrial sectors to include environmental protection in their management and production processes. The Green Council is a non-profit, tax-exempt charitable environmental stewardship organisation and certification body (Reg. No.: HKCAS-027) of Hong Kong established in 2000. A group of individuals from different sectors of industry and academics shared the vision to help build Hong Kong into a world-class green city for the future. They formed the Green Council with the aim of encouraging the commercial and industrial sectors to include environmental protection in their management and production processes. The Green Council is a non-profit, tax-exempt charitable environmental stewardship organisation and certification body (Reg. No.: HKCAS-027) of Hong Kong established in 2000. A group of individuals from different sectors of industry and academics shared the vision to help build Hong Kong into a world-class green city for the future. They formed the Green Council with the aim of encouraging the commercial and industrial sectors to include environmental protection in their management and production processes.

Contact:
Website: https://www.greencouncil.org/hkgls
Email: info@greencouncil.org
Telephone: (852) 2810 1122

BIO LATINA

CERTIFICADORA

Peru

BIO LATINA, the consolidated byproduct of four Latin American national certification entities.Since 1998, we have provided certification services in Latin America for national and international markets. We seek to help create a more sustainable and resilient world. With these goals in mind, we have expanded our service portfolio beyond organic to social and environmental certifications.

Visit us: https://biolatina.com

From our regional offices we serve Latin American.

Our headquaters:
Av. Javier Prado Oeste 2501, Bloom Tower Of. 802, Magdalena del Mar,
 Lima 17, Perú

sustainable florist

Netherland

For more than 25 years, the independent Dutch foundation SMK works from professional knowledge with companies to improve the sustainability of products and business management. SMK cooperates with an extensive stakeholder network of governments, producers, branch and non-governmental organisations, retailers, consultancies, researchers. The SMK Boards of Experts establish objective criteria for more sustainable products and services. SMK's transparent work processes, third party audits and certifications are conducted according to international certification standards, mostly under supervision of the Dutch Accreditation Council. Besides, SMK is Competent Body of the EU Ecolabel. SMK keeps an extensive database of sustainability criteria.

Contact:
Bezuidenhoutseweg 105 - 2594 AC Den Haag
Telefoon: 070-3586300
Mobiel: 06-82311031
(niet op woensdag)
www.smk.nl

Korea
Eco-Label

Republic of Korea

The Korea Eco-labelling is a certification system enforced by the Ministry of Environment and KEITI(Korea Environmental Industry & Technology Institute). Since its foundation in April 1992, the system has certified a wide range of eco-friendly products, which were selected as excellent not only in terms of their environmental-friendliness, but also for their quality and performance during their life cycle. Korea Eco-labelling is voluntary certification scheme to attach logo to products with superior environmental quality throughout their lifecycle to other products of the same use, and thus to provide product information to consumers. For 30 years, the scheme has launched plenty of eco-labelling product standards covering personal and household goods, construction materials, office equipment furniture, etc. It products categories which cover all aspects of products, such as reduction of use of harmful substances, energy saving, resource saving, etc. As of April 30th 2021, 169 criterias(=standards), and certifications for 18,250 products(4,549 companies) have maintained.

Contact:
Korea Environmental Industry & Technology Institute(KEITI)
Office of Korea Eco-Label Innovation
Address: 215, Jinheung-ro, Eunpyeong-gu, Seoul, Repulic of Korea
T: +82 2 2284 1518
F: +82 2 2284 1526
E: accolly@keiti.re.kr
W: www.keiti.re.kr

USA

The Carbonfree® Product Certification is a meaningful, transparent way for you to provide environmentally-responsible, carbon neutral products to your customers. By determining a product's carbon footprint, reducing it where possible and offsetting remaining emissions through our third-party validated carbon reduction projects, companies can:
•   Differentiate their brand and product
•   Increase sales and market share
•   Improve customer loyalty
•   Strengthen corporate social responsibility & environmental goals

The Carbonfree® Product Certification Program is proud to be part of Amazon's Climate Pledge Friendly Program!
Carbonfund.org is leading the fight against climate change, making it easy and affordable to reduce & offset climate impact and hasten the transition to a clean energy future.

Contact:

O: 240.247.0630 ext 633
C: 203.257.7808
M: 853 Main Street, East Aurora, NY, 14052

Taiwan

The Green Mark GM) Program was launched by the Environmental Protection Administration of Taiwan (TEPA) in 1992. As the official Type I eco-labeling program, it is in compliance with the requirements of the international stadard, ISO 14024 and is considered an important tool to promote green consumption and production .

To improve the GM application/review mechanism and introduce a third party certification scheme, TEPA promulgated the «Guideline for the Management of Certification Organizations for Environmental Protection Products" in June 2012. Both Environment and Development Foundation (EDF) and the Taiwan Testing and Certification Center (ETC) were commissioned by TEPA as official certifiers. With the expansion of certification capacity and authorization of the certification decision, the certification time was greatly reduced.

Contact :

Website: www.edf.org.tw
TEL: 886-3-5910008 #39
E-mail: lhliu@edf.org.tw

Denmark, Finland, Norway, Iceland, Sweden

The Nordic Swan Ecolabel

The Nordic Swan Ecolabel is the official Nordic ecolabel supported by all Nordic Governments. It is among the world›s strictest and most recognised environmental certifications.

The Nordic Swan Ecolabel is a Type I environmental labelling program established in 1989 by the Nordic Council of Ministers, connect¬ing policy, people, and businesses with the mission to make it easy to make the environmentally best choice. Nordic Ecolabelling is the non-profit organisation responsible for the Nordic Swan Ecolabel.

The organisation offers independent third-party certification and support for a wide range of product areas and services, ensuring that they comply with the Nordic Swan Ecolabel's strict requirements through documentation and inspections.

30 years of experience and expertise has made the Nordic Swan Ecolabel a powerful tool that paves the way to a sustainable future by giving producers a recipe on how to develop more environmentally sustainable products, and giving consumers credible guidance by helping them identify products that are among the environmentally best.

Globally, you can find more than 25,000 Nordic Swan ecolabelled products. 93% of all Nordic consumers recognise the Nordic Swan Ecolabel as a brand, and 74% believe that the Nordic Swan Ecolabel makes it easier for them to make envi¬ronmentally friendly choices (IPSOS 2019).

Denmark, Finland, Norway, Iceland, Sweden

Securing a sustainable future

The Nordic Swan Ecolabel works to reduce the overall environmental impact from production and consumption and contributes significantly to UN Sustainable Development Goal 12: Responsible consumption and production.

To ensure maximum environmental impact, the Nordic Swan Ecolabel sets product specific requirements and evaluates the environmental impact of a product in all relevant stages of a product lifecycle - from raw materials, production, and use, to waste, re-use and recycling.

Common to all products certified with the Nordic Swan Ecolabel is that they meet strict environmental and health requirements. All requirements must be documented and are verified by Nordic Ecolabelling. Nordic Ecolabelling regularly reviews and tightens the requirements.

Therefore, certifications are time-limited and companies must re-apply to ensure sustainable development.

International website:

Nordic-ecolabel.org

National websites:

Denmark: ecolabel.dk

Sweden: svanen.se

Norway: svanemerket.no (in Norwegian)

Finland: joutsenmerkki.fi (in Finnish)

Iceland: svanurinn.is   (in Icelandic)

EU Ecolabel
www.ecolabel.eu

EUROPE

Established in 1992 and recognized across Europe and worldwide, the EU Ecolabel is a label of environmental excellence that is awarded to products and services meeting high environmental standards throughout their life-cycle: from raw material extraction, to production, distribution and disposal. The EU Ecolabel promotes the circular economy by encouraging producers to generate less waste and CO_2 during the manufacturing process. The EU Ecolabel criteria also encourages companies to develop products that are durable, easy to repair and recycle.

The EU Ecolabel criteria provide exigent guidelines for companies looking to lower their environmental impact and guarantee the efficiency of their environmental actions through third party controls. Furthermore, many companies turn to the EU Ecolabel criteria for guidance on eco-friendly best practices when developing their product lines. The EU Ecolabel helps you identify products and services that have a reduced environmental impact throughout their life cycle, from the extraction of raw material through to production, use and disposal. Recognised throughout Europe, EU Ecolabel is a voluntary label promoting environmental excellence which can be trusted.

Spain , Germany, Italy, Sweden, Greece, Portugal, Poland, Belgium, Netherlands, Estonia, Finland, Austria, Lithuania, Czech Republic, Norway, Cyprus, Ireland, Slovenia, Hungary, Romania, Croatia, Bulgaria, Malta, Slovak Republic, Latvia, Luxembourg, Iceland

Contact and more information via: http://ec.europe.eu

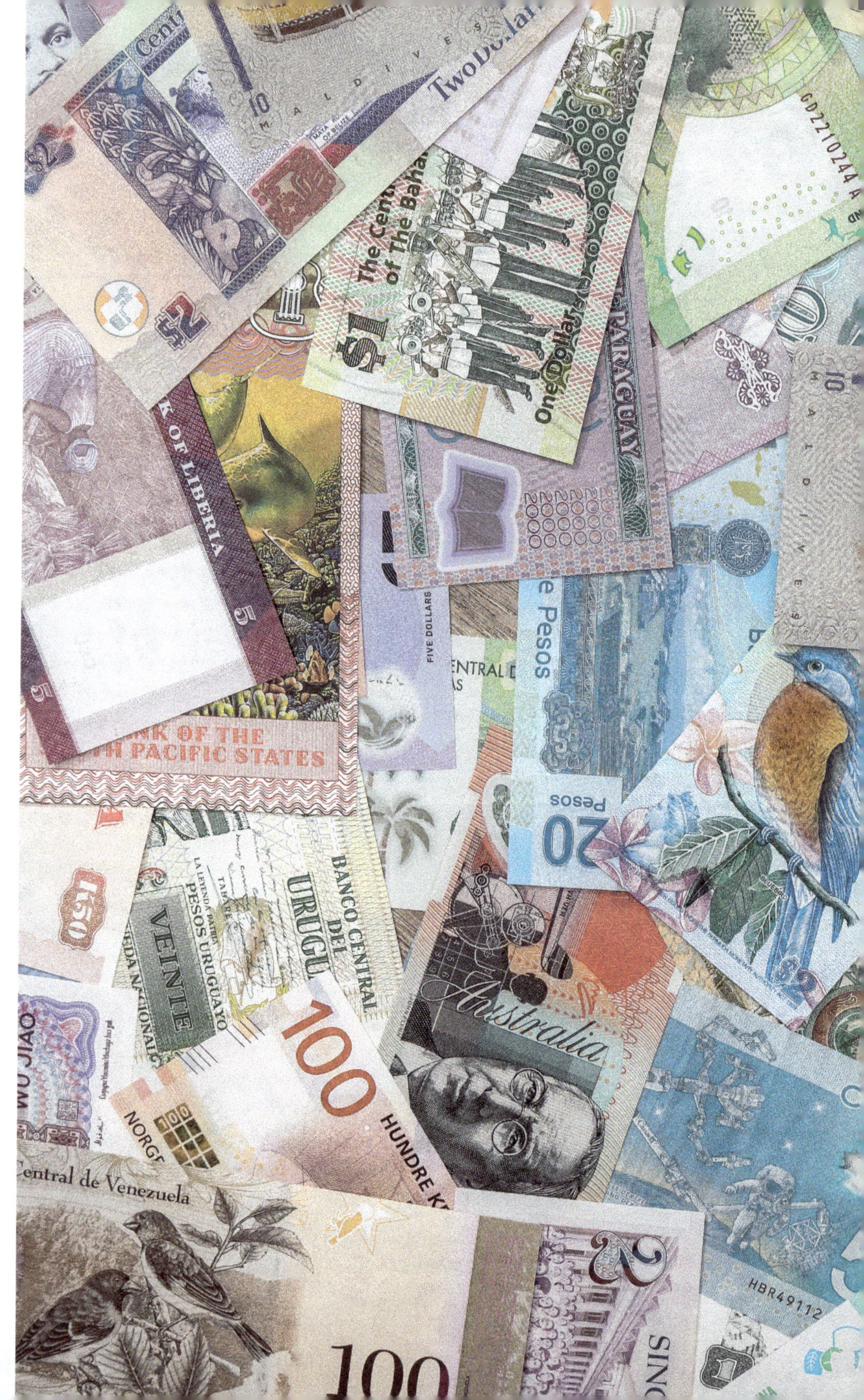

DOS MIL PESOS
Bank of Papua New Guinea
Two Kina
RESERVA DEL PERU
ZAIRE
5 OCTOBER 2015
500
QUINIENTOS
RESERVE BANK
200
500
NGA MAG
20
SCHWEIZERISCHE
BANCA NAZIUNALA
BU172
بینک دولت پاکستان
OF JORDAN
BANCO
NGAN HA
DONG
OF ERITREA
BANCO D
100
ARIARY

CHAPTER 5

Type II Environmental Labelling

Type II environmental labelling refers to the claims made on product labels in connection with business centers. This includes familiar claims such as recyclable, ozone-free, 60% phosphate-free, and the like. This type of labelling can be in the form of a mark or sentence on the product packaging. Some of them are valid environmental claims—and some can be completely misleading.

Usually, all countries have laws against deceptive advertisements, so why has the International Organization for Standardization discussed this issue? The answer is that it is not clear whether the environmental claims have a technical basis or whether the ad is meaningless.

Most countries have guidelines at the national level to help producers and consumers know what constitutes a true, scientifically valid claim.

There is a national standard on this in Canada. In Australia, the Consumer Commission has published guidance on this, and there are similar examples in other countries.

Canada

Environmental Sustain for Future kids established in Vancouver, BC Canada in 2020. (ESFK) is an international ecolabel focused on taking care of environment for future of kids.

ESFK defined as 'self-declared' environmental claims made by manufacturers and businesses based on ISO 14020 series of standards, the claimant can declare the environmental objectives and targets in relation to taking care of environment for future kids. However, this declaration will be verifiable.

Environmental Sustain for Future Kids
Vancouver, BC CANADA

Email: info@esfk.org
Web: www.esfk.org

What is eco friendly notebook?

The 11 Most Sustainable Cryptocurrencies:
SolarCoin (SLR)
Powerledger (POWR)
Cardano (ADA)
Stellar (XLM)
Nano (NANO)
IOTA (MIOTA)
EOSIO (EOS)
TRON (TRX)
Signum (SIGNA)
Holochain/HoloTokens (HOT)
DEVVIO

CHAPTER **6**

Type III Environmental Labelling

T ype III environmental labelling is a distinct form of third-party environmental labelling pattern designed to avoid the difficulties that can result from type I labelling. Technical committee for Environment of International organization for Standardization has undertaken a new project to standardize guidelines and Type III labelling methods. One of the main objections raised by industries to Type I labelling is the basis for its management.

Due to the nature of the system, less than 50% of the various products on the market can meet the criteria and qualify for Type I Labelling. As long as the industry is the main supporter of other third-party models for quality systems, it is sometimes difficult for an industry to support a program that can only benefit 15% of its members. This type of labelling is currently practiced in some countries, such as Sweden, Canada, and the United States. Choosing the right product has never been easy, but Type III labelling will help because each product can have a label that describes its environmental performance and is certified by a third-party company. Consumers can then compare labels and choose their favorite products.

CHAPTER 7

All about 'Eco-friendly' Financial Products and/or Services

Green loans, energy efficiency mortgages, alternative energy venture capital, eco-savings deposits, and "green" credit cards; these items represent merely a handful of innovative, "green" financial products that are currently offered around the globe1 . In an age where environmental risks and opportunities abound, so too have the options for reconciling environmental matters with lending and financing arrangements.

The purpose of this Volume is to examine the currently available Eco friendly financial products and services, with a focus on lesson learning opportunities, the nature and transferability of best practices, and how key designs can potentially increase market share and generate profits, while improving brand recognition and enhancing reputation.

In this volume we have divides the financial services sector into the following categories:

1) Green Retail Banking;
2) Green Corporate and Investment Banking;
3) Insurance
4) Asset Management;

1. Green Retail Banking

Retail banking covers personal and business banking products and services designed for individuals, households and SMEs, rather than large corporate or institutional clients. Products and services in the eco friendly retail space include Green loans and mortgages, debit and credit card services, travelers' cheques, money orders, overdraft protection, cash management services and insurance, among others.

2. Green Corporate and Investment Banking

Corporate and investment banking, or "wholesale banking", sees banks provide banking solutions to large corporations, institutions, governments and other public entities with complex financial needs, typically international in scope. Financial institutions offering corporate and investment banking can underwrite debt issues, both on their own behalf and for corporate and public sector clients, as well as supply equity, manage funds and offer advice to corporate mergers and acquisitions. These banks act as financial intermediaries, raising capital (equity and debt) by trading foreign exchange, commodities and equity securities on the primary market.

3) Insurance

The insurance sector can generally be divided into two categories:
Life Insurance; and General (Non-Life) Insurance.
"Green" insurance falls under the latter and typically encompasses two product areas:
1) those which allow an insurance premium differentiation on the basis of environmentally relevant characteristics; and
2) Insurance products specifically tailored for clean technologies and emissions reducing activities.

4) Asset Management

Asset Management has become one of the fastest growing segments in the financial industry and represents a core business unit of current banks. This space focuses on providing financial advice to clients on estate planning, mutual funds, managed asset programs, taxes, trust services, international financial planning, global private banking and full-service and discount brokerages.

Green Retail Banking

Ask something more from your local Bank

Home Mortgage	• 'green' mortgage initiative. %1 reduction on interest for loans that meet environmental criteria. • Free home energy rating and offsets carbon emissions for every year of loan • Green Power Oriented Mortgage • 10% premium refund on its mortgage loan insurance
Commercial Building Loan	• Green Loans for new condos • Developers do not have to pay an initial premium for "green" commercial buildings • Provides 1/8 of 1% discount on loans to green leadership projects
Auto Loan	Clean Air Auto Loan with preferential rates for hybrids & Electrical
Credit Card	Climate Credit Card. Bank will donate to WWF GreenCard Visa is the world's first credit card to offer an emissions offset program
Deposit	Fully-insured deposits earmarked for lending to local energy-efficient companies aiming to reduce waste/pollution, or conserve natural resources

Green Corporate and Investment Banking

Ask something more from your local Bank

FINANCE Project Finance	Specialized service divisions are dedicated to long-term financing of clean energy projects. Some banks also specialize in one (or several) renewable technology type and/or place a premium on working with states where regulatory framework and government policy encourages the early adoption of clean technologies.
Securitization	A risk sharing arrangement for environmental projects. Financial institution represents a guarantor (or structuring investor) at the mezzanine level of risk, allowing client to transfer risk to bank. Eco-Securitization scheme will test the feasibility of financing "natural infrastructure" by linking sustainable management of resources with the funding capacity and requirements of asset-backed securitization.
Carbon Finance	Banks provide equity, loans and/or upfront or upon delivery payments to acquire carbon credits from projects. Most acquire carbon credits in order to serve their corporate clients' compliance needs, supply a tradable product to the banks' trading desks, or develop lending products backed by emission allowances and carbon credits.
Indices	Series of environmental private investor eco-market products includes a biofuels commodity basket, total returns solar energy index, clean renewable energy index and total returns water index (e.g., enables interested parties to invest in water as a commodity).
Deposit	Fully-insured deposits earmarked for lending to local energy-efficient companies aiming to reduce waste/pollution, or conserve natural resources

Insurance

Auto Insurance	Pay As You Drive™ Insurance. Mileage-based Insurance. 10% discount for hybrid and fuel efficient vehicles. Bank can also choose to offset vehicle's annual emissions (e.g. 20 % emissions offset through Climate Care Recycling Insurance. Customer pays less for car insurance, by up to 20%, if recycled parts are used when vehicle is damaged and requires service.
Building and Home Insurance	Green Building Replacement and Upgrade Coverage. Product covers unique type of "green" risks related to the sustainable building industry. "Climate Neutral" Home Insurance Policy. First home insurance product to carry out GHG offsetting based on customer usage.
Business Insurance	Environmental Damage Insurance.

Asset Management

Fisacl Green Fund	By purchasing shares or investing in Green Funds, customers receive an income tax discount, and thus accept a lower interest rate on investment. Banks can offer loans at lower cost to finance environmental projects related to eligible categories.
Fund	Eco Performance is the world's largest "green" fund. %80 of assets are channeled towards eco and social leaders, with %20 going to "eco-innovators". Equity Fund - Future Energy, focuses on clean energy sector investments in clean four energy-related business segments.
Cat Bond Fund	Cat Bond Fund. World's first public fund for catastrophe bonds, a portion of which is aimed at climate-related natural disasters (or climate adaptation). Vehicle designed to hedge climate risks typically difficult to cover in the traditional insurance market.

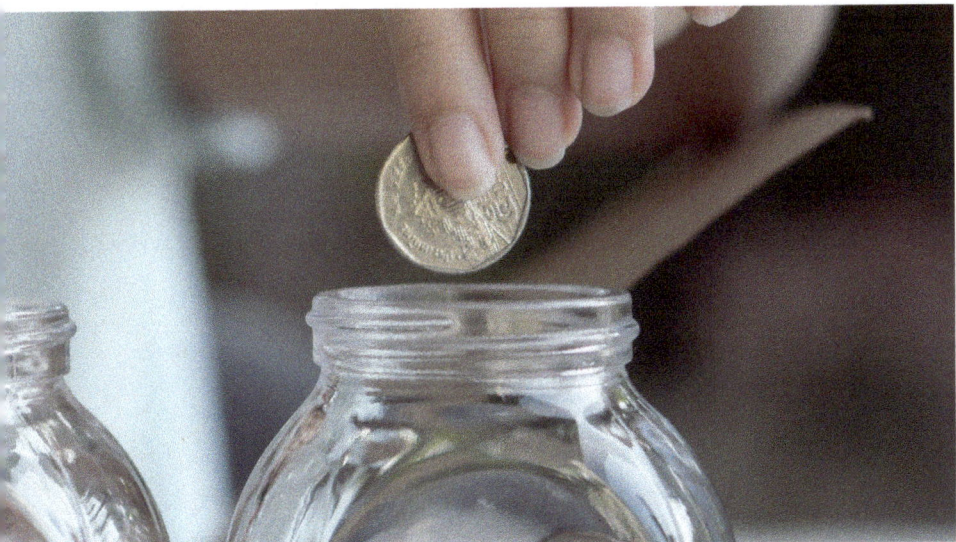

Conclusion

Many "green" financial products and services, reviewed above, either remain in the nascent stage of development/implementation or data related to their success/failure has not yet been generated or reported.

Due to this lack of experience and data, any rigorous measurement or ranking of these designs would be overly speculative and risk misrepresenting some designs over others. Looking ahead, however, as more quantitative and qualitative track records emerge for these products,

The following questions should be considered when gauging product performance or promise:

- Does it achieve high levels of financial performance?
- Does it attract a particularly large number of customers?
- Does it last over time, and is re-launched year by year?
- Does it raise the environmental awareness among all stakeholders, including clients and employees?
- Does it receive positive attention from the media and environmental NGOs?
- Does it prompt the introduction of other environmental products and services?
- Does it improve brand recognition and corporate image among stakeholders?

As environmental understanding and awareness grows in North America, so too will the demand for products and services aimed at facilitating the advancement of environmentally sustainable lives, livelihoods and communities. At the same time, this demand will also expose new business opportunities, while leading to an increased diversification of products and services found in multiple sectors. Consequently, organizations that have the foresight and capacity to tap into this desire by consumers to affect positive environmental change will likely experience widespread benefits; from improved corporate image to increased growth and competitiveness in the marketplace. Given their intermediary role in the economy and farreaching customer base, financial institutions will be well-positioned to reap financial and non-financial rewards, while furthering their contribution toward sustainable development.

Are Crypto Currencies Inherently Bad For The Environment?

All cryptocurrencies have an energy and environmental problem. But if done right, it might be possible to channel all that energy into something good for the planet.

Crypto's environmental troubles

A fierce debate around the environmental impacts of cryptocurrencies, like bitcoin, is growing. Bitcoin does consume a lot of energy. That energy use is growing and annually consumes as much energy as whole nations, such as Finland, Malaysia, or Sweden. While bitcoin is not the only industry to consume as much energy as entire countries, e.g. concrete consumes more energy than India, the energy both sectors consume comes with associated pollution, including carbon emissions.

Even transactions with bitcoin use a lot of energy, with the average transaction consuming over 1,700 kWh of electricity, which is almost twice the monthly amount used by the average U.S. home. However, there was ways to transact in bitcoin using much less energy.

Exacerbating this problem, some bitcoin mining operations have teamed up with struggling fossil fuel power plants, keeping some power plants online that would otherwise have retired, increasing overall carbon emissions. Some utilities have even gotten into the bitcoin game directly.

Large bitcoin mining operations are also moving locations as China, the country previously with the largest bitcoin mining industry, recently banned both cryptocurrency mining and transactions. This change has bitcoin mining operations moving to places like Texas and potentially Alberta, Canada.

All else equal, bitcoin operations that co-locate and utilize fossil fuels that would have otherwise stayed in the ground will increase emissions.

Some are considering using stranded natural gas that would otherwise have been flared, which, absent any methane venting and flaring regulations, would make the use of the natural gas for bitcoin, at-best, carbon neutral. However, it is a stretch, and making the natural gas more valuable at the wellhead could further dissuade pipeline development that would have moved the gas to market.

However, co-locating bitcoin mining operations with zero-carbon resources, such as nuclear, hydro, wind and solar, could help reduce the carbon emissions associated with the mining itself. Co-location could also give a financial boost to power plants that might be able to sell their electricity at a higher price to miners instead of to the grid when demand and prices are low. This type of hybrid power plant/mine might even make uneconomical projects economical.

Going further, it is also possible that the cryptocurrency mines themselves could offer benefits directly to the grid, and, if operated intelligently, even result in lower overall carbon emissions.

A positive grid impact?

The study simulated the evolution of the Electric Reliability Council of Texas (ERCOT), the grid that serves most of Texas, out to 2030 under multiple scenarios:
1) a base case with no datacenter/bitcoin mining expansion,
2) a case with 5 GW of inflexible (always on) datacenter/bitcoin mines by 2030,
3) a scenario with 5 GW of mildly flexible datacenters deployed by 2030 and
4) a scenario with 5 GW of very flexible datacenters deployed by 2030.

The non-flexible scenario added a significant baseload to the ERCOT system. This growth resulted in the deployment of more power plant capacity that the base case, including more wind, natural gas, and solar.
This increased energy use also resulted in an additional 7.9 million metric tons of carbon emissions over the base case by 2030.

However, the flexible scenarios were more interesting. Both flexible scenarios actually see more wind and less natural gas deployed than both the base case and the inflexible scenarios.

This change is because the datacenters/mines were programed to reduce their energy consumption by certain percentages when electricity prices hit certain tiers. In total, the third scenario saw the datacenters/mines curtailing their load about 14% of the year.

The flexibility of the datacenters/mines in the latter two scenarios allowed the model to deploy different levels of technologies than the base or inflexible case. The model actually built more renewables because it could utilize the flexibility of the datacenters/mines to compensate for fluctuations in renewable output. This flexibility also resulted in lower carbon emissions compared to the base case.

For additional load to result in lower total carbon emissions, the additional energy consumption must be offset by more zero-carbon energy. In the flexible datacenter/mine cases, the amount of energy generated from wind and solar was more than in the base case and the amount generated from natural gas was lower.

In general, the flexibility of the datacenters/mines moves their load to more value energy over power, which better aligns with renewables. This is because renewables are great at providing large amounts of energy, but have less ability to always provide capacity, or constant power.

In concept, flexible datacenters/mines are similar to the electrification of transportation or heating with the ability to control then times when the chargers and heaters operate. However, it is likely that datacenters/mines could offer large levels of flexible load concentrated in a smaller number of locations, which could make their administration easier.

Grid decarbonization studies often assume high levels of flexible demand, and often much of this flexibility comes from diffuse sources, such as smart thermostats and EV charging. While this analysis did not seek to satisfy any carbon policy, it does illustrate the potential carbon benefits of high levels of flexible demand coupled with an electricity market that is able to incorporate it.

Mining and transacting cryptocurrencies, such as bitcoin, do present energy and emissions challenges, but new research shows that there are possible pathways to mitigate some of these issues if cryptocurrency miners are willing to operate in a way to compliment the deployment of more low-carbon energy.

The author of this Book does not currently own or mine any cryptocurrencies.

Garibaldi Lake, Near Whistler, , British Columbia, Canada

Consumers are increasingly demanding sustainable products that are not toxic to themselves or the environment. The natural market is growing exponentially, and choosing raw, natural materials will cement your brand as a safe choice — both environmentally and economically

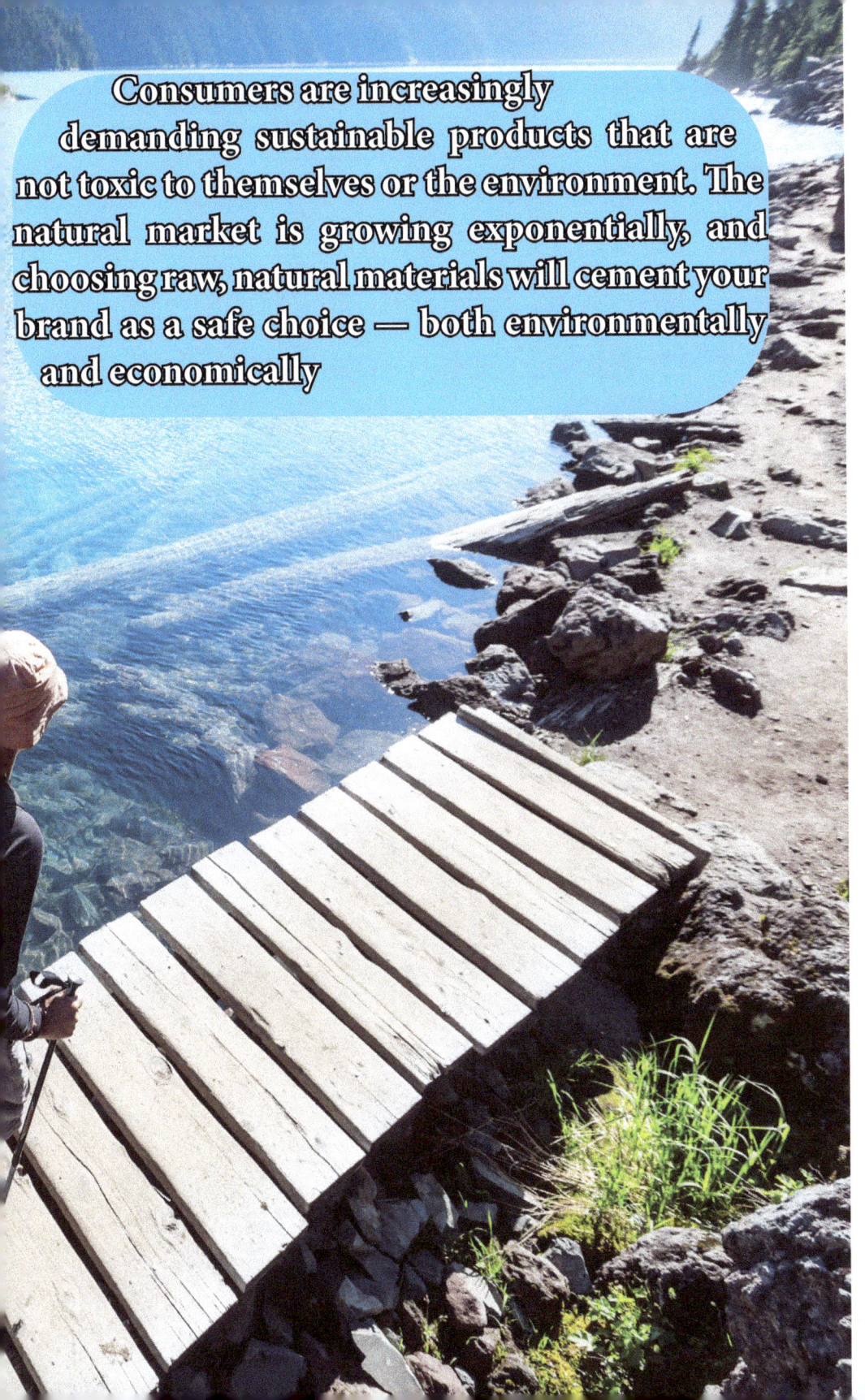

CHAPTER 9

Top Ten Award International Network Environmental Pioneers

T op Ten Award international Network (TTAIN) was established in 2012 to recognize outstanding individuals, groups, companies, organizations representing the best in the public works profession. TTAIN publishing books related to international Eco-labeling plans to increase public knowledge in purchasing based on the environmental impacts of products. We introduce in each volume some of the organizations that are doing their best in relation to taking care of the environmnet.

FSC **FORESTS FOR ALL FOREVER**

Germany

FSC® is a global not-for-profit organization that sets the standards for responsibly managed forests, both environmentally and socially. When timber leaves an FSC certified forest they ensure companies along the supply chain meet our best practice standards also, so that when a product bears the FSC logo, you can be sure it's been made from responsible sources. In this way, FSC certification helps forests remain thriving environments for generations to come, by helping you make ethical and responsible choices at your local supermarket, bookstore, furniture retailer, and beyond. www.fsc.org

FSC® International
Adenauerallee 134
53113 Bonn
E-mail: info@fsc.org
Phone: +49 (0) 228 367 66

FSC Canada
50 rue Sainte-Catherine Ouest,
bureau 380B, Montreal, QC H2X 3V4
Email: info@ca.fsc.org
Telephone: 514-394-1137

Note:
We've done our absolute best to provide the best information possible, but since we haven't tried every single one of these solutions in every possible cleaning situation, we can't vouch for them 100 percent.

UN environment programme

UNEP

The United Nations Environment Programme (UNEP) is the leading global environmental authority that sets the global environmental agenda, promotes the coherent implementation of the environmental dimension of sustainable development within the United Nations system, and serves as an authoritative advocate for the global environment.

Our mission is to provide leadership and encourage partnership in caring for the environment by inspiring, informing, and enabling nations and peoples to improve their quality of life without compromising that of future generations.

Headquartered in Nairobi, Kenya, we work through our divisions as well as our regional, liaison and out-posted offices and a growing network of collaborating centres of excellence. We also host several environmental conventions, secretariats and inter-agency coordinating bodies. UN Environment is led by our Executive Director.

We categorize our work into seven broad thematic areas: climate change, disasters and conflicts, ecosystem management, environmental governance, chemicals and waste, resource efficiency, and environment under review. In all of our work, we maintain our overarching commitment to sustainability.

Website: www.unep.org

Bibliography:

Amberg, N.; Magda, R. Environmental Pollution and Sustainability or the Impact of the Environmentally Conscious Measures of International Cosmetic Companies on Purchasing Organic Cosmetics. Visegrad J. Bioecon. Sustain. Dev. 2018, 1, 23.

Asadi, J., "International Environmental Labelling, Economic Consequencies, Export Magazine, July 2001

Asadi, J. 2008. Mobile Phone as management systems tools, ISO Magazine, Vol.8, No.1

Asadi, J., Eco-Labelling Standards, National Standard Magazine, Sep. 2004.

Barbieux, D.; Padula, A.D. Paths and Challenges of New Technologies: The Case of Nanotechnology-Based Cosmetics Development in Brazil. Adm. Sci. 2018, 8, 16.

Advanced Engineering and Applied Sciences: An International Journal 2014; 4(3): 26-28

Berolzheimer, C. (2006). Pencils: An Environmental Profile.

Chemical Week, 1999. Europe's Beef Ban Tests Precautionary Principle. (August 11).

Chaudri, S.K.; Jain, N.K. History of Cosmetics. Asian J. Pharm. 2009, 7–9, 164–167.

CHOI, J.P. Brand Extension as Informational Leverage. Review of Eco- nomic Studies, Vol. 65 (1998), pp. 655-669.

Conway, G. 2000. Genetically modified crops: risks and promise.

Corrado, M., (1989), The Greening Consumer in Britain, MORI, London

Corrado, M., (1997), Green Behaviour – Sustainable Trends, Sustainable Lives?, MORI, london, accessed via countries. Manila, Asian Development Bank 33p.

Davies, Clive. Chief, Design for the Environment Program, Environmental Protection Agency. Interview. March 24, 2009.

Federal Trade Commission, "Sorting Out Green Advertising Claims." http://www.ftc. gov/bcp/edu/pubs/consumer/general/gen02.shtm (March 26, 2009, March 27, 2009)

Ooyen, Carla. Research Manager with Nutrition Business Journal. Personal correspondence. March 19, 2009.

Tekin, Jenn. Marketing Manager with Packaged Facts & SBI. Personal correspondence. March 17, 2009.

University of California - Berkeley. http://berkeley.edu/news/media/releases/2006/05/22_ householdchemicals.shtml (March 26, 2009)

U.S. Department of Health and Human Services, Household Products Database.http:// householdproducts.nlm.nih.gov/cgi-bin/household/prodtree?prodcat=Inside+the+Home (March 17,

Women's Voices of the Earth, "Household Cleaning Products and Effects on Human Health."http://www.womenandenvironment.org/campaignsandprograms/SafeCleaning/safecleaninghealth (March 17, 2009)

ISO 14020, ISO 14021,ISO 14024,ISO 14025, International Organization for Standardization.

Environmental Finance. 2007(a). In Brief: FTSE4Good Sets Out Climate Criteria. March.

Environmental Finance. 2007(b). Corporate Profile: New Resource Bank (NRB) - Making Their Green, Greener. February.

Environmental Finance. 2007 (c). ABN Amro Launches Climate Change Index. March 29. http://www.environmental-finance.com/onlinews/0329abn.htm

Environmental Finance. 2003. Are Cat Bonds Changing Course? http://www.environmental-finance.com/2003/0304apr/catbonds.htm

Esty, D. and A. Winston. 2006. Green to Gold: How Companies Use Environmental Strategy to Innovate, Create Value and Build Competitive Advantage. New Haven.

Current Trends and Future Opportunities in North America 51

ETA (Environmental Transportation Agency). 2007. Britain's First Climate Neutral Home Insurance to Offset Customers' Usage. Press Release. February 1. www.eta.co.uk

EuroMoney Institutional Investor, 2005. http://www.invenergyllc.com/

Fireman's Fund. 2007. Green-Gard Coverages. www.firemansfund.com

Funding Universe. Corporation History: MBNA. http://www.fundinguniverse.com

GreenBiz.com. 2006(a). New Resource Bank Announces Plan to Finance Residential Solar Systems. October 24. San Francisco. www.greenbiz.com

GreenBiz.com. 2006(b). Survey: Majority of Investment Managers Link Corporate Responsibility to Asset Performance.

Green Finance Products and Services, NATF Report, AUgust 2007

March 17. http://www.greenbiz.com/news/news_third.cfm?NewsID=30617

GreenBiz.com. 2004. Builders Break Ground on World's Most Environmentally Responsible High Rise Office Building. August 4. www.greenbiz.com

Greener Buildings. 2007. New Bank Aims to Make It Easier to Build Green. Produced by GreenBiz.com. February 1. http://www.greenbiz.com/sites/greenerbuildings/news_detail.cfm?NewsID=34525

GreenCard Visa. 2007. Netherlands. http://www.greencardvisa.nl/

Green Car Congress. 2007. Reported US Sales of Hybrids Up 26% in April. May 3. http://www.greencarcongress.com/2007/05/reported_us_sal.html

Grundon. 2005. Financial Close for Slough Energy from Waste Facility Announced by Lakeside Energy from Waste Ltd.News Bulletin. October 3. http://www.grundon.com

Harris, A. 2005. Supporting Climate Change Solutions. VanCity Presentation for Pollution Probe on Complementary

Mechanisms. June 28. www.pollutionprobe.org/Happening/pdfs/complementarymeas-June28/harris.pdf

IISD (International Institute for Sustainable Development). 2007. Business and Sustainable Development: A Global Guide. www.bsdglobal.com

Innis, R. 2006. A Theory of Consumer Boycotts under Symmetric Information and Imperfect Competition. Economic Journal. Vol. 116, issue 511, pg 355-381.

Institute for Market Transformation to Sustainability. 2005. Special Report: Green Mortgage-Backed Securities One Step Closer to Reality. Building Operating Management Magazine. December. http://www.facilitiesnet.com/bom/article.asp?id=3623

ISO (International Standards Organization). 1999. Financing Catastrophe Risk: Capital Market Solutions.. Studies and Analyses. http://www.iso.com/studies_analyses/hurricane_experience/financingrisk.html#14

Labatt, S. and R.W. White. 2007. Carbon Finance: The Financial Implications of Climate Change. John Wiley & Sons Inc.

Liu, Peter. Founder and CEO of New Resource Bank (NRB). Interview conducted in February 2007.

Joshua D. Rhodes, Ph.D., Is Bitcoin Inherently Bad For The Environment?

Jeucken, M. 2004. Sustainability in Finance: Banking on the Planet.

Jeucken, M. 2001. Sustainable Finance and Banking – Slow starters are gaining pace. Based on Jeucken's book Sustainable Finance and Banking. 2001. www.sustainability-in-finance.com

Kennedy, P.E. "Estimation with Correctly Interpreted Dummy Variables in Semilogarithmic Equations," American Economic Review 71: 801 (1981).

Kirchho®, S., (2000), Green Business and Blue Angels.

Kraus, Jeff. Lab Technician at the North Carolina School of Textiles.

Labeling Issues, Policies and Practices Worldwide.

Lamport, L. 1998. The cast of (timber) certifiers: who are they? International J. Ecoforestry 11(4): 118-122.

Large Scale impoverishment of Amazonian forests by logging and fire. 1999.

Lathrop, K.W. and Centner, T.J. 1998. Eco-labeling and ISO 14000: An analysis of US regulatory systems and issues concerning adoption of type II standards. Environmental

Lee, J. et al. 1996. Trade related environmental measures; sizing and comparing impacts.

Lehtonen, Markku. 1997. Criteria in Environmental Labeling: A comparative Analysis on Environmental Criteria in Selected Labeling Schemes. Geneva, UNEP. 148p.

LIEBI, T. Trusting Labels: A Matter of Numbers? Working Paper Uni versity of Bern, No. 0201 (2002).

Lindstrom, T. 1999. Forest Certification: The View from Europe's NIPFs. Journal of Forestry 97(3): 25-31. London

Losey, J.E., Rayor, L.S. & Carter, M.E. 1999. Transgenic pollen harms monarch larvae. Nature 399 20 May): p.214.

Mattel Ever After High Cedar Wood Doll. (2014, July 3).

Management 22 (2) : 163-172.

Mattoo, A. and H. V. Singh, (1994), Eco-Labelling: Policy Considera-Michaels, R. G., and V. K. Smith. "Market Segmentation And Valuing Amenities With Hedonic Models: The Case Of Hazardous Waste Sites," Journal of Urban Economics, 1990 28(2), 223-242.

Nicholson-Lord, D., (1993) 'Tis the Season to be Green, The Independent, 20 December

Nuttall, N., (1993), Shoppers can cross green products off their lists, The Times, 3 July

OCDE/GD(97)105. Paris, OECD. 81p.

OECD. "Ec-labelling: Actual Effects of Selected Programmes," OCDE/GD (97) 105, 1997, Paris. (available on line at http://www.oecd.org/env/eco/books.htm#trademono)

OECD. 1997a. Case study on eco-labeling schemes. Paris, OECD (30 Dec):

OECD. 1997b. Eco-labeling: Actual Effects of Selected Programs.

Osborne, L. "Market Structure, Hedonic Models, and the Valuation of Environmental Amenities." Unpublished Ph.D. dissertation. North Carolina State University, 1995.

Osborne, L., and V. K. Smith. "Environmental Amenities, Product Differentiation, and market Power," Mimeo, 1997.

Ozanne, L.K. and Vlosky, R.P. 1996. Wood products environmental certification: the United States perspective". Forestry Chronicle 72 (2) : 157-165.

Palmquist, R. B., F. M. Roka, and T.Vukina. "Hog Operations, Environmental Effects, and Residential Property Values," Land Economics 73(1), (1997): 114-24.

Palmquist, R.B. "Hedonic Methods," in J.B Braden and C.D. Kolstad, eds. Measuring the Demand for Environmental Improvement. Amsterdam, NL: Elsevier, 1991.

Paper Mate. (2014). Paper Mate Recycled.

Pento, T. 1997. Implementation of Public Green Procurement Programs (22-31) in Greener Purchasing: Opportunities and Innovations. Sheffield, Greenleaf Publ. 325 p.

Perloff, J. "Industrial Organization Lecture Notes," Mimeo. University of California at Berkeley (1985).

Plant, C. and Plant, J. 1991. Green business: hope or hoax? Philadelphia, New Society Publishers 136 p.

Pencil Making Today (2014, January 1). Pencil Making Today: How to Make a Pencil in 10 Steps.

Polak, J. and Bergholm, K. 1997. Eco-labeling and trade: a cooperative approach (Jan.): Policy in a Green Market. Environmental and Resource Economics 22, 419-

Poore, M.E.D. et al. 1989. No timber without trees. London, Earthscan. 352p.

Raff, D. M.G., and M. Trajtenberg. "Quality-Adjusted Prices for the American Automobile Industry: 1906-1940." NBER Working Paper Series, Working Paper No. 5035, February 1995.

Roberts, J. T. 1998. Emerging global environment standards: prospects and perils. Journal of Developing Societies 14 (1): 144-163.

Rosen, S., "Hedonic Prices and Implicit Markets: Product Differentiation in Pure Competition." Journal of Political Economy. 82: 34-55 (1974).

Ross, B. 1997. Eco-friendly procurement training course for UN HCR. : 126 p.

Ryan, S., and Skipworth, M., (1993), Consumers turn their backs on green revolution, The Times, 4 April

Salzman, J. 1997. Informing the Green Consumer: The Debate over the Use and Abuse of Environmental Labels. Journal of Industrial Ecology 1 (2): 11-22.

Sanders, W. 1997. Environmentally Preferable Purchasing: The US Experience (946-960) in Greener Purchasing: Opportunities and Innovations. Sheffield, Greenleaf Publ. 325p.

Sayre, D. 1996. Inside ISO 14000: The competitive advantage of environmental management. Delray Beach FL., St. Lucie Press. 232p.

Suzuki, D. (2014, January 1). PEG Compounds and their contaminants

SHAPIRO, C. Premiums for High Quality Products as Returns to Reputa- tion. Quarterly Journal of Economics, Vol. 98, No. 4 (1983), pp. 659-680.

Stillwell, M. and van Dyke, B. 1999. An activists handbook on genetically modified organisms and the WTO. Washington DC., The Consumer's Choice Council: 20 p.

Semenzato, A.; Costantini, A.; Meloni, M.; Maramaldi, G.; Meneghin, M.; Baratto, G. Formulating O/W Emulsions with Plant-Based Actives: A Stability Challenge for an Eective Product. Cosmetics 2018, 5, 59.

Sources of Plastics (2014, January 1). Sources of Plastics.

Singh, S. (2008, March 6). Paraffin wax.

Saint Jean Carbon. (n.d.). Sri Lankan Graphite.

Teisl, M. F., B. Roe, and R. L. Hicks. "Can Eco-labels tune a market? Evidence from dolphin-safe labeling," Presented paper at the 1997 American Agricultural Economics Association Meetings, Toronto.

Tollefson, Jennifer E. (2008). Calocedrus Decurrens.

THE GERSEN, C. Psychological Determinants of Paying Attention to Eco- Labels in Purchase Decisions: Model Development and Multinational Vali- dation. Journal of Consumer Policy, Vol. 23, No. 4 (2000), pp. 285-313.

Tibor, T. and Feldman, I. 1995. ISO 14000: a guide to the new environmental management standards. Burr Ridge Ill., Irwin Professional Publ. 250 p.

TU.S. Energy Information Administration, What is U.S. Electricity Generation by Energy Source?, Retrieved From: https://www.eia.gov/tools/faqs/faq.php?id=427&t=3

U.S. Energy Information Administration, Biomass Explained, Retrieved From: https://www.eia.gov/energyexplained/?page=biomass_home

U.S. Environmental Protection Agency. National Water Quality Fact Inventory: 1990 Report to Congress. EPA 503-9-92-006, Apr. 1992.

UK Eco-labelling Board website, accessed via http://www.ecosite.co.uk/Ecolabel-UK/

US Environmental Protection Agency (EPA742-R-99-001): 40 p. <www.epa.gov/opptintr/epp>

US EPA, 1993. Determinants of effectiveness for environmental certification and labeling programs. Washington, D.C., US Environmental Protect

US EPA, 1993. Status report on the use of environmental labels worldwide. Washington, D.C., US Environmental Protection Agency (742-R-93-001 September).

US EPA, 1993. The use of life-cycle assessment in environmental labeling. Washington, D.C., US Environmental Protection Agency (742-R-93-003 September).

US EPA, 1998. Environmental labeling: issues, policies, and practices worldwide. Washington DC., Environmental Protection Agency, Pollution Prevention Division Prepared by Abt

US EPA, 1999. Comprehensive procurement guidelines (CPG) program. Washington, D.C., US Environmental Protection Agency: <www.epa.gov/cpg>

US EPA, 1999. Environmentally preferable purchasing program: Private sector pioneers: How companies are incorporating environmentally preferable purchases. Washington University of Saskatchewan, Sustainable purchasing guide.

USG, 1993. Federal acquisition, recycling, and waste prevention. Washington DC., Executive Order: (20 October).

USG, 1998. Greening the government through waste prevention, recycling, and federal acquisition. Washington, D.C., Executive Order 13101 (September).

Kijjoa, A.; Sawangwong, P. Drugs and Cosmetics from the Sea. Mar. Drugs 2004, 2, 73–82. [CrossRef]

Wang, J.; Pan, L.; Wu, S.; Lu, L.; Xu, Y.; Zhu, Y.; Guo, M.; Zhuang, S. Recent Advances on Endocrine Disrupting Eects of UV Filters. Int. J. Environ. Res. Public Health 2016, 13, 782.

Bilal, A.I.; Tilahun, Z.; Shimels, T.; Gelan, Y.B.; Osman, E.D. Cosmetics Utilization Practice in Jigjiga Town, Eastern Ethiopia: A Community Based Cross-Sectional Study. Cosmetics 2016, 3, 40.

Ting, C.T.; Hsieh, C.M.; Chang, H.-P.; Chen, H.-S. Environmental Consciousness and Green Customer Behavior: The Moderating Roles of Incentive Mechanisms. Sustainability 2019, 11, 819.

Chen, K.; Deng, T. Research on the Green Purchase Intentions from the Perspective of Product Knowledge. Sustainability 2016, 8, 943.

Wang, H.; Ma, B.; Bai, R. How Does Green Product Knowledge Eectively Promote Green Purchase Intention? Sustainability 2019, 11, 1193.

Nguyen, T.T.H.; Yang, Z.; Nguyen, N.; Johnson, L.W.; Cao, T.K. Greenwash and Green Purchase Intention: The Mediating Role of Green Skepticism. Sustainability 2019, 11, 2653.

Cinelli, P.; Coltelli, M.B.; Signori, F.; Morganti, P.; Lazzeri, A. Cosmetic Packaging to Save the Environment: Future Perspectives. Cosmetics 2019, 6, 26.

Eixarch, H.; Wyness, L.; Siband, M. The Regulation of Personalized Cosmetics in the EU. Cosmetics 2019, 6, 29.

APPENDIX I: SEARCH BY LOGOS

H ere you can search the logos in this volume. It will help you to better undersand the Ecolabels you may encounter while shopping. Buying Eco-products will aid in having a better environment with minimum polution during production processes. Three important parameteres for shopping are **quality**, **price** & **environmental impacts** of the products.

Vol.6 Goto page: 31	Vol.6 Goto page: 42

Vol.6 Goto page: 48	Vol.6 Goto page: 29

Vol.6 Goto page: 36	Vol.6 Goto page: 34

Vol.6 Goto page: 29	Vol.6 Goto page: 43

Vol.6　Goto page:　35	Vol.6　Goto page:　37
Vol.6　Goto page:　48	Vol.6　Goto page:　29
Vol.6　Goto page:　29	Vol.6　Goto page:　32
Vol.6　Goto page:　40 ,41	Vol.6　Goto page:　43

Vol.6 Goto page: 48	Vol.6 Goto page: 30
Vol.6 Goto page: 39	Vol.6 Goto page: 34
Vol.6 Goto page: 31	Vol.6 Goto page: 38
Vol.6 Goto page: 33	Vol.6 Goto page: 29

Appendix II

Federal Investments in Algae Cultivation Technologies

There is a reason that algae is poised to make an outsized impact on consumer markets, low-carbon energy and sustainable solutions to global challenges: forward-looking investments in advanced algal technology from the public sector.

While there's no doubt that venture capital and strategic investments from major companies fuel much growth in the industry, it's important to recognize the critical role that government funding has played (and still does play) in fueling this sector's growth.

For decades, agencies like the Departments of Energy and Agriculture, the National Science Foundation and others have been tasked by Congress to support the long-term innovations that will help build a more prosperous America.

It quickly became apparent that those technologies would have applications in food, animal nutrition and other markets that could be disrupted by the availability of a more sustainable, fast-growing, and nutrient-packed crop.

In the decade since, the federal government's investments in algae research and commercialization have led to transformational technology advances; game-changing product innovations; and, by our estimate, over $1 billion in private investments in U.S.-based research and development, agriculture and manufacturing infrastructure, and jobs.

APPENDIX III

Environmental Friendly Photos

E nvironmental friendly photos will be placed in this appendix. These photos can be received in the Top Ten Award International Network inbox from anywhere and everywhere, all over the globe. You can send your appropriate photos to us for them to be considered for publishing in one of the future, related volumes. They will be published with proper credit to the sender. The pictures can also be images of the Ecolabels existing in products within your country.

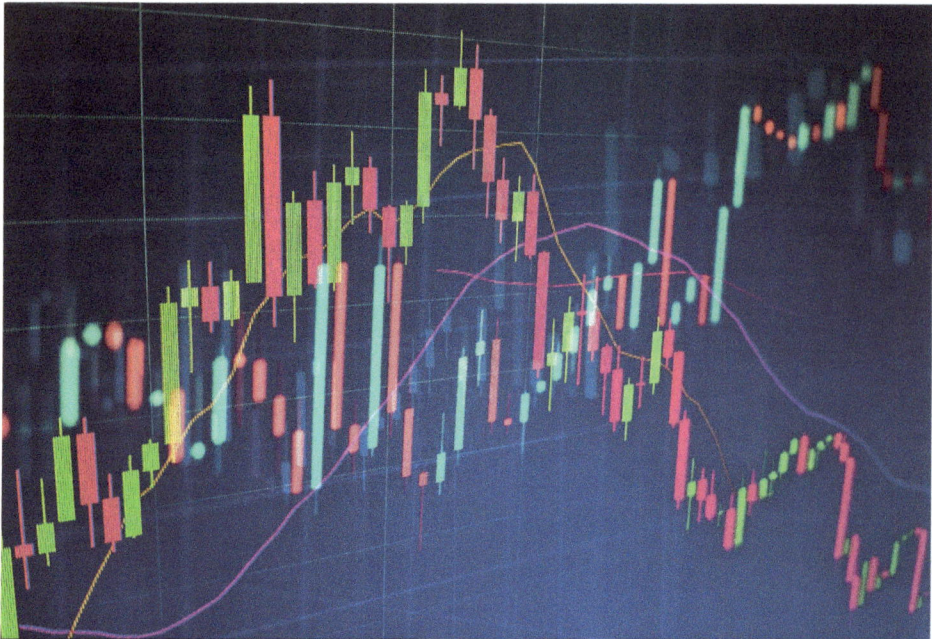

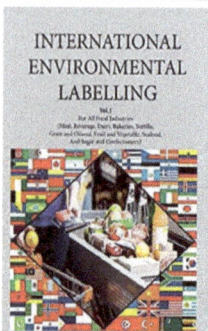

INTERNATIONAL ENVIRONMENTAL LABELLING Vol.1 For All Food Industries (Meat, Beverage, Dairy, Bakeries, Tortilla, Grain and Oilseed, Fruit and Vegetable, Seafood, And Sugar and Confectionery)	# Vol.1 For All People who wish to take care of Climate Change, Food Industries: (Meat, Beverage, Dairy, Bakeries, Tortilla, Grain and Oilseed, Fruit and Vegetable, Seafood, And Sugar and Confectionery)
INTERNATIONAL ENVIRONMENTAL LABELLING Vol.2 For All Energy & Electrical Industries (Renewable Energy, Biofuels, Solar Heating & Cooling, Hydroelectric Power, Solar Power, Wind Power, Energy Conservation, Geothermal and Nuclear Power) JAHANGIR ASADI	# Vol.2 For All People who wish to take care of Climate Change, Electrical Industries: (Renewable Energy, Biofuels, Solar Heating & Cooling, Hydroelectric Power, Solar Power, Wind Power, Energy Conservation, Geothermal and Nuclear Power)
INTERNATIONAL ENVIRONMENTAL LABELLING Vol.3 For All Fashion & Textile Industries (Fashion Design, The Fashion System, Fashion Retailing, Marketing and Merchandizing, Textile Design and Production, Clothing and Textile Recycling) JAHANGIR ASADI	# Vol.3 For All People who wish to take care of Climate Change, Fashion & Textile Industries: (Fashion Design, The Fashion System, Fashion Retailing, Marketing and Marchandizing, Textile Design and Production, Clothing and Textile Recycling)
INTERNATIONAL ENVIRONMENTAL LABELLING Vol.4 For All Health & Beauty Industries (Fragrances, Makeup, Cosmetics, Personal Care, Sunscreen, Toothpaste, Bathing, Nailcare & Shaving, Skin Care, Foot Care, Hair Care and Other Health & Beauty Products) JAHANGIR ASADI	# Vol.4 For All People who wish to take care of Climate Change, Health & Beauty Industries: (Fragrances, Makeup, Cosmetics, Personal Care, Sunscreen, Toothpaste, Bathing, Nailcare & Shaving, Skin Care, Foot Care, Hair Care and Other Health & Beauty Products)

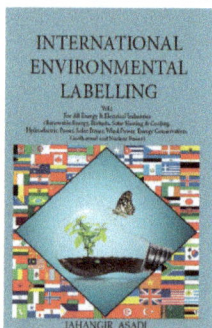

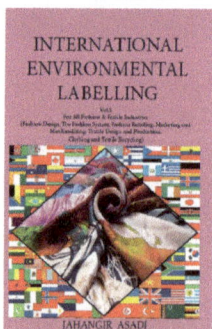

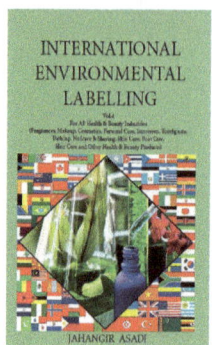

INTERNATIONAL ENVIRONMENTAL LABELLING Vol.5 For All Maintenance & Cleaning Products All-purpose Cleaners, Abrasive Cleaners, Powders, Liquids, Specialty Cleaners, Kitchen, Bathroom, Glass and Metal Cleaners, Bleaches, Disinfectants and Disinfectant Cleaners) JAHANGIR ASADI	**Vol.5** For All People who wish to take care of Climate Change, Maintenance & Cleaning Products: (All-purpose Cleaners, Abrasive Cleaners, Powders. Liquids, Specialty Cleaners, Kitchen, Bathroom, Glass and Metal Cleaners, Bleaches, Disinfectants and Disinfectant Cleaners)
INTERNATIONAL ENVIRONMENTAL LABELLING Vol.6 (Wooden Products, Cardboard, Papers, Markers, Pens, NoteBooks, Writing Pads and Writing Sets, Pencils, White Papers, Envelopes and Organizers, Staplers and Paper Clips) JAHANGIR ASADI	**Vol.6** For All People who wish to take care of Climate Change, Wood & Stationery Industries: (Wooden Products, Cardboard, Papers, Markers, Pens, NoteBooks, Writing Pads and Writing Sets, Pencils, White Papers, Envelopes and Organizers, Staplers and Paper Clips)
INTERNATIONAL ENVIRONMENTAL LABELLING Vol.7 JAHANGIR ASADI	**Vol.7** For All People who wish to take care of Climate Change, DIY & Construction Industries: (Do it yourself " ("DIY") of Building, Modifying, or Repairing, Renovation, Construction Materials, Cement, Coarse Aggregates. Clay Bricks, Power Cables, Pipes and Fittings, Plywood, Tiles, Natural Flooring)
INTERNATIONAL ENVIRONMENTAL LABELLING Vol.8 JAHANGIR ASADI	**Vol.8** For All People who wish to take care of Climate Change, Agricuture & Gardening Industries: (Shifting Cultivation, Nomadic Herding, Livestock Ranching, Commercial Plantations, Mixed Farming, Horticulture, Butterfly Gardens, Container Gardening, Demonstration Gardens, Organic Gardening)

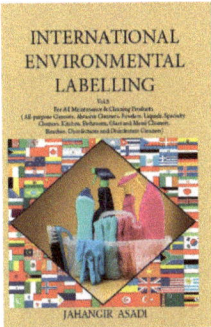

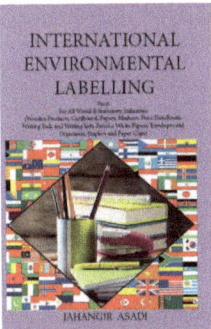

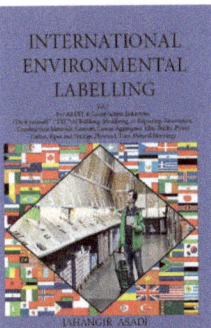

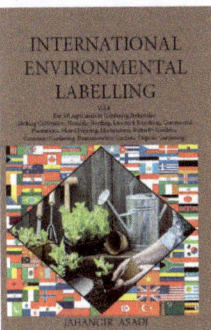

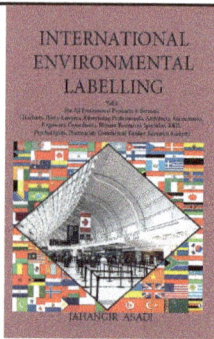

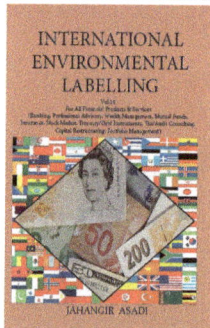

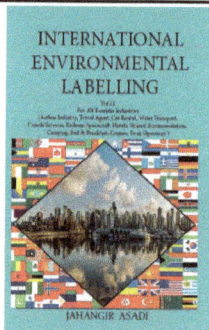

www.ingramcontent.com/pod-product-compliance
Lightning Source LLC
Chambersburg PA
CBHW040757220326
41597CB00029BB/4976